AF509999

Table of Contents

1 Introduction to Malware

Malware, in its simplest form, is any software intentionally designed to cause damage, disruption, or unauthorized access to a computer system. Understanding this fundamental definition is crucial, as it encompasses a vast and ever-evolving landscape of threats. From the relatively simple viruses of the past to the highly sophisticated and adaptable malware of today, the common thread is the malicious intent behind their creation. This intent can range from financial gain through ransomware to espionage and the disruption of critical infrastructure. The impact, however, transcends simple technical damage; it often involves significant financial losses, data breaches, reputational harm, and even physical harm in extreme cases. The consequences necessitate a deep understanding of this malicious software's workings, which is precisely the goal of this chapter.

The sheer variety of malware types is staggering. Viruses, often likened to biological counterparts, require a host program to replicate and spread. They infect files and can trigger various actions, ranging from subtle data alteration to complete system crashes. Worms, unlike viruses, are self-replicating and spread autonomously across networks, often exploiting vulnerabilities to gain entry and replicate themselves. Their ability to proliferate rapidly can lead to widespread network disruptions and data loss. Trojan horses, deceptively disguised as legitimate software, often gain entry through user interaction, installing malicious code that executes without the user's knowledge. This often allows for data exfiltration, remote system control, or the installation of further malicious software. Spyware secretly monitors user activity, often stealing sensitive information such as passwords, banking details, or personal data. Ransomware, a particularly prevalent threat, encrypts data and demands a ransom for its release, often resulting in significant financial losses for individuals and organizations alike. Each type possesses unique characteristics and deployment methods, requiring a tailored approach for detection and mitigation.

Understanding the mechanisms behind each malware type is paramount. Viruses often use techniques like code injection to embed themselves within legitimate programs. Worms leverage network protocols and vulnerabilities to propagate, often targeting weak points in system security. Trojans employ social engineering tactics, luring unsuspecting users into executing malicious code. Spyware often hides its presence using rootkit techniques, making detection extremely difficult. Ransomware often uses strong encryption algorithms to render data inaccessible, maximizing its impact. These diverse methods highlight the complexity and sophistication of modern malware and underscore the need for proactive security measures.

Beyond the individual types, malware often operates as part of larger, coordinated attacks. Botnets, networks of compromised computers controlled remotely by attackers, are frequently used to launch distributed denial-of-service (DDoS) attacks, overwhelming targeted servers and rendering them inaccessible. These attacks can cripple online services, disrupt businesses, and even target critical infrastructure. The scale and sophistication of these botnets highlight the collaborative and organized nature of many cybercriminal operations. Furthermore, sophisticated techniques like polymorphism and metamorphism allow malware to adapt and evade detection, making them particularly difficult to neutralize. These techniques involve constantly altering the malware's code without changing its functionality, rendering signature-based detection methods ineffective.

Effective malware analysis is crucial in understanding its behavior, identifying its origins, and developing countermeasures. This process usually involves both static and dynamic analysis. Static analysis entails scrutinizing the malware's code without actually executing it, attempting to identify suspicious functions or patterns. This method offers a high-level overview of the malware's capabilities and potential threats but may miss behaviors that only manifest during runtime. Dynamic analysis, conversely, involves executing the malware in a controlled environment, typically within a virtual machine or sandbox. This allows security professionals to observe the malware's behavior in real-time, identifying its actions, network connections, and potential targets. This approach provides a more comprehensive understanding of the malware's functionalities, but it requires specialized tools and expertise, and it also presents a risk of malware escaping the controlled environment if precautions are not properly taken. A comprehensive approach often combines both techniques for a complete picture.

Ultimately, understanding the nature of malware, its different forms, and the methods used to analyze it is critical for anyone involved in cybersecurity. This foundation is essential for developing effective prevention strategies, incident response plans, and proactive measures to protect against the ever-evolving threat of malicious software. The constant evolution of malware necessitates ongoing education, adaptation, and refinement of security practices, highlighting the ongoing arms race between attackers and defenders in the ever-shifting digital landscape. The proactive approach is crucial – waiting for an attack to happen is often too late.

1.1 Defining Malware

Malware, in its simplest form, is any software designed to damage, disrupt, or gain unauthorized access to a computer system. It's a broad term encompassing a wide range of malicious programs, each with its own unique characteristics and goals. Understanding this fundamental definition is crucial before delving into the complexities of specific malware types. This is not simply a technical definition, but one that emphasizes the destructive intent behind these programs, a common thread that unites them despite their varied forms.

Unlike benign software, malware operates surreptitiously, often concealing itself within seemingly harmless files or exploiting vulnerabilities in operating systems and applications. This stealthy behavior is a key characteristic, differentiating it from software that openly declares its purpose and functionality. The insidious nature of malware necessitates proactive security measures and a thorough understanding of its methods.

The objectives of malware creators vary widely, ranging from financial gain to espionage and sabotage. Some malware aims to encrypt a user's data and demand a ransom for its release, a tactic known as ransomware. Others may seek to steal sensitive information, such as passwords or credit card details, for identity theft or financial fraud. Still others might be designed to disrupt services or cause widespread damage, potentially impacting critical infrastructure. The motivations behind these malicious programs are as diverse as the programs themselves.

Identifying malware often requires a detailed analysis of its behavior and characteristics. Unlike legitimate software, malware rarely displays its malicious intent openly. Instead, it typically employs deceptive techniques to avoid detection, masking its true purpose. Recognizing these deceptive practices is a crucial skill for anyone working in cybersecurity. The subtle differences between legitimate processes and those of malware are the key to its detection.

Furthermore, the evolution of malware has mirrored advancements in technology and security practices. Early malware was relatively simple, often consisting of self-replicating code with limited capabilities. Modern malware, however, frequently incorporates sophisticated techniques such as polymorphism, allowing it to change its code structure to evade detection by antivirus software. This constant evolution necessitates a dynamic approach to security, requiring ongoing adaptation and innovation.

Analyzing the code and the behavior of malware can provide valuable insights into its functionality, origin, and purpose. This analysis helps determine the type of malware involved, enabling a better response to the threat. By studying how malware operates, security professionals develop more effective countermeasures and strategies to prevent future attacks. This constant cycle of action and adaptation is an integral part of cybersecurity.

The impact of malware extends far beyond individual users. Large-scale malware attacks can cripple critical infrastructure, disrupting essential services such as power grids, hospitals, and financial institutions. The consequences of such attacks can be devastating, both economically and socially. The implications of even a small breach can be far-reaching, impacting individuals and nations alike. This widespread impact underscores the importance of robust security measures at all levels.

Effective countermeasures often involve a multi-layered approach. This may include implementing firewalls to protect network perimeters, deploying intrusion detection systems to identify suspicious activity, and utilizing endpoint security software to secure individual devices. However, a crucial aspect often overlooked is the importance of user education. By raising awareness among users about common malware threats and practices, organizations can significantly reduce the risk of infection. A well-informed user is the first line of defense against malware.

Understanding the underlying principles of how malware functions, its various forms, and its objectives, and how security strategies are implemented helps lay the groundwork for more effective cybersecurity practices. By comprehending the strategies used by malware, one can start to develop a proactive approach to security, not just a reactive one. It's not enough to merely react to threats; it's crucial to anticipate and prevent them. The knowledge presented here provides a crucial starting point on this path, forming the foundation for more advanced topics to be explored in subsequent chapters.

1.2 Types of Malware

Viruses, the oldest and most pervasive type, are self-replicating programs that attach themselves to other files. They spread by infecting executable files or documents, often requiring user interaction, such as opening an infected email attachment. Their primary goal is often to replicate and spread, sometimes causing damage as a side effect, though increasingly, viruses are used as components in more complex malware suites, their ability to spread rapidly providing a powerful vector for more insidious threats. The damage can range from minor system slowdowns to complete data loss. Understanding their basic propagation mechanisms is critical, as even seemingly benign viruses can act as a foothold for more devastating attacks. The development of anti-virus software significantly mitigated their direct impact, yet their use as building blocks in modern malware remains relevant.

Worms, unlike viruses, are self-contained programs that can replicate and spread independently without needing to attach to other files. They exploit network vulnerabilities to spread rapidly, often causing significant disruption by consuming bandwidth or overloading systems. The notorious "Morris Worm" in 1988 serves as a prime example, demonstrating the potential for catastrophic network-wide disruption caused by a seemingly simple program. While the initial shock of widespread worm infestations has lessened, they remain a threat, particularly to poorly secured networks and devices; their ability to spread autonomously makes them a constant concern for network administrators. Focusing on patching known vulnerabilities and implementing robust network segmentation strategies is crucial in minimizing their impact.

Trojans, deceptively disguised as legitimate software, hide malicious code within seemingly harmless applications. Users unknowingly install them, granting access to their systems. Once installed, Trojans can perform various malicious activities, from stealing data to installing further malware. Their insidious nature lies in their ability to bypass user suspicion, acting as a silent backdoor into compromised systems. This makes early detection exceptionally difficult and highlights the importance of only downloading software from trusted sources and regularly updating security software. Understanding common Trojan disguises, such as fake security software or cracked games, is a vital step towards preventing infection.

Ransomware, a particularly damaging type, encrypts a victim's files, rendering them inaccessible until a ransom is paid. Often spread through phishing emails or exploit kits, ransomware presents a significant financial and operational threat, impacting businesses and individuals alike. The sophistication of ransomware has increased dramatically over time, with attackers demanding increasingly higher ransoms and employing advanced encryption techniques that make decryption challenging, even with specialized tools. Effective preventative measures, such as robust backups, regular security updates, and employee training on identifying phishing attempts, are essential in mitigating the risk of ransomware attacks. Understanding the mechanics of encryption used by ransomware and the tactics used to deliver it are also invaluable in building a resilient defense.

Spyware, designed to monitor user activity and steal sensitive information, operates surreptitiously in the background. It can track browsing history, keystrokes, and other personal data, often transmitted to a remote server controlled by the attacker. Unlike ransomware, spyware's impact may not be immediately apparent, but the long-term consequences of data theft can be severe. Understanding the various methods spyware uses to collect and transmit data, such as through hidden processes or compromised browser extensions, is crucial for detection and prevention. A strong emphasis on protecting one's online privacy, including the use of strong passwords, secure web browsing habits, and regular security software updates, is paramount in preventing spyware infection.

Rootkits, highly advanced malware, gain privileged system access, making them exceptionally difficult to detect. They conceal their presence and operations, making them particularly challenging to remove. Their ability to manipulate the system at the lowest level is what sets rootkits apart from other malware types. Often used in conjunction with other malware to maintain persistent access, rootkits can enable attackers to perform various malicious actions without detection. The development of advanced detection techniques, as well as a strong focus on system hardening and regular security audits, are crucial in defending against rootkits. Knowing how rootkits function, including their use of system hooks and kernel-level manipulation, is fundamental to understanding their threat.

Adware, though less overtly malicious than other types, displays unwanted advertisements and can track user browsing habits. While not directly harmful, adware can compromise system performance, and the data collected can be used for targeted advertising or even more insidious purposes. Often bundled with free software or downloaded from untrusted websites, adware presents a persistent challenge in maintaining a clean and efficient computing environment. Being discerning about software downloads and installing ad blockers can significantly reduce adware infections. Understanding the methods adware uses to intrude and the implications of data collection related to adware is important for informed user decisions.

2 Early Malware Evolution

The earliest forms of malware were surprisingly simple. Think of them as digital pranks, more mischievous than truly malicious. The first computer viruses, appearing in the 1970s, were often self-replicating programs designed to spread through floppy disks, the primary method of data transfer at the time. They weren't particularly sophisticated, typically displaying a simple message or slightly altering the computer's behavior – a far cry from the complex threats we see today. These early efforts, however, laid the foundation for the more dangerous malware that would follow.
Creeper, one of the earliest known computer viruses, could replicate itself and travel across ARPANET, the precursor to the internet. It wasn't destructive, but it demonstrated the potential for self-propagating code to spread uncontrollably. This early warning proved largely unheeded, paving the way for the rapid evolution and increasing sophistication of malware. The sheer novelty of the concept, coupled with the relative infancy of computer security, meant that these initial threats often went unnoticed or were dismissed as mere curiosities. This lack of awareness would have significant consequences in the years to come.
The next significant step in malware evolution was the emergence of worms. Unlike viruses, which require a host program to infect, worms are self-contained programs capable of independent replica-

tion and spreading across networks. The Morris Worm, released in 1988, is a prime example. This worm exploited vulnerabilities in the Unix operating system, causing widespread network disruption and demonstrating the potential for widespread damage through network propagation. This marked a crucial turning point. The Morris worm's ability to exploit vulnerabilities and spread autonomously highlighted a critical shift in the threat landscape. It was no longer just about infecting individual computers; it was about disrupting entire networks. The scale of the disruption caused by the Morris Worm served as a stark wake-up call to the growing vulnerability of interconnected systems.

Trojan horses, unlike viruses and worms, don't replicate themselves. Instead, they disguise themselves as legitimate software, enticing users to install them. Once installed, Trojans can perform a variety of malicious actions, ranging from stealing sensitive information to granting remote access to the infected system. Their deceptive nature and ability to bypass user awareness made them highly effective, and arguably even more dangerous than earlier malware. The insidious nature of Trojan horses highlighted the importance of user vigilance and the limitations of technical security measures alone. This underscored a shift towards a more holistic approach to cybersecurity, incorporating human factors into the overall strategy.

The transition from simple self-replicating programs to sophisticated malware capable of exploiting vulnerabilities and evading security measures was remarkably swift. This rapid evolution was driven by several factors, including the increasing complexity of computer systems, the expansion of networks, and the growing sophistication of malicious actors. The early days of malware development, while simplistic compared to modern threats, laid the groundwork for the complex and damaging attacks we see today.

The early malware landscape was a breeding ground for innovation, albeit of a malicious nature. Each new form of malware built upon the successes and failures of its predecessors. The lessons learned, both by attackers and defenders, shaped the future of the cybersecurity landscape. The seemingly innocuous beginnings of these early threats eventually paved the way for the far more dangerous and widespread problems we grapple with today. Understanding this evolution is crucial to appreciating the ongoing threat, and to developing effective defense strategies. The relatively simple techniques used in these early days provided a template for future attacks, showing the inherent vulnerability of emerging technologies to malicious exploitation. This early experience underscores the importance of proactive security measures rather than merely reactive responses to evolving threats. We must constantly learn from the past, lest we repeat its mistakes.

The evolution of malware didn't happen in isolation. Technological advancements, such as the proliferation of the internet and the increased reliance on networked systems, fuelled this evolution. This created a fertile ground for the spread of malicious code, increasing the potential impact of even relatively simple programs. Understanding this interplay between technological advancements and the evolution of malicious code is crucial for anticipating and mitigating future threats. The increasingly interconnected nature of our digital world only amplifies the potential damage of malware attacks.

Early malware may seem primitive compared to today's sophisticated threats, yet they represent a crucial chapter in the ongoing battle between cybercriminals and security professionals. The fundamental principles of self-replication, network propagation, and social engineering remain central to many modern attacks. Studying these early forms of malware gives us valuable insights into the strategies and tactics used by malicious actors, allowing us to anticipate future threats and develop more effective defenses. The continuous adaptation of malware reinforces the necessity for ongoing vigilance and the constant evolution of security measures to stay ahead of the curve. The foundation of modern cyber threats was built on the seemingly simple beginnings of these early malware programs, creating a timeline of escalating complexity.

The lessons learned from analyzing early malware are invaluable.

They highlight the importance of robust security practices, such as regular software updates, strong passwords, and user education. The simplicity of early malware, ironically, made them powerful learning tools. They exposed fundamental vulnerabilities and highlighted the need for proactive measures. The early history of malware serves as a constant reminder of the ever-evolving nature of cybersecurity threats and the necessity for vigilance. The constant arms race between attackers and defenders demands continuous innovation and adaptation in the face of the ever-changing threat landscape. Every step forward in security technology is met by an equally rapid adaptation from those seeking to exploit vulnerabilities. The past informs the future, and understanding the past is critical to confronting the present and anticipating what's to come.

2.1 The Genesis of Viruses

The earliest forms of computer viruses weren't the sophisticated, network-spanning threats we see today. They were simpler, often residing within single machines and spreading through physical means, like floppy disks. Think of them as digital hitchhikers, passively waiting to be transported to new hosts. Their code, usually written in assembly language, was remarkably small, sometimes only a few hundred lines, demonstrating the efficiency born from necessity in those early days of computing. These weren't malicious in the grand scheme of things, but more akin to pranks or tests of skill amongst programmers, demonstrating the power of manipulating code within the infancy of the field. Their simplicity belied their significance, however, as they represented the genesis of a technological arms race—the first shots fired in a conflict that continues to this day.

One of the earliest known examples, the "Creeper" program, was not technically a virus by today's definition; it simply replicated itself across the ARPANET, the precursor to the internet. It wasn't destructive, but its self-replication capability served as a foundational proof-of-concept for the viruses that would follow. This innocent act of experimentation paved the way for more malicious actors to leverage the same techniques and create more harmful programs. Its message, a simple "I'm the creeper, catch me if you can," served as a surprisingly apt description of the future of malware: always evolving, always seeking to evade capture.

Subsequent viruses demonstrated a growing sophistication. They began to incorporate destructive capabilities, erasing data or corrupting files, escalating the stakes of this digital game. This shift marked the transition from playful experimentation to something far more sinister. The evolution wasn't linear, however. There were periods of rapid advancement, punctuated by pauses, perhaps as developers explored new techniques or struggled with the limitations of the hardware and software of that era. The focus shifted from simple replication to inflicting tangible damage, demonstrating a clear escalation in malevolence.

The emergence of the internet greatly accelerated the evolution of viruses. Suddenly, the possibility of global spread became a reality, transforming isolated incidents into widespread epidemics. Network connectivity allowed viruses to spread exponentially faster and further than ever before. The sheer number of potential targets, coupled with the ease of propagation, represented a paradigm shift in the way viruses operated. No longer confined to a single machine or a local area network, the potential impact of even a relatively simple virus was amplified dramatically. This dramatic shift forced the development of new countermeasures, highlighting the dynamic nature of the ongoing arms race.

Researchers and security experts began developing anti-virus software to combat this escalating threat. This early software, often simple signature-based systems, attempted to identify known viruses through code matching. The 'cat-and-mouse' game began: Virus writers found ways to evade detection, and security researchers responded with improved techniques. The early battles were often won and lost within this constant adaptation and counter-adaptation. These were simple beginnings, but they laid the foundation for the complex, multi-layered security systems that we rely on today.

This constant back-and-forth created an environment where innovation and creativity were rewarded. The most successful virus writers became masters of obfuscation, encoding their creations to make them harder to analyze and identify. This forced the security community to develop more sophisticated analysis techniques, such as dynamic analysis and reverse engineering. The process, although complex and seemingly endless, shaped the field of cybersecurity as we know it, driving significant improvements in code analysis, network protection, and security protocols.

Furthermore, the very nature of viruses—their reliance on host systems for propagation—led to a deep understanding of how software operates. Studying viruses often revealed unexpected vulnerabilities in operating systems and applications, leading to improvements in software design and security practices. In essence, the study of viruses pushed the boundaries of software engineering and cybersecurity, forcing ongoing improvements in both offense and defense. The indirect benefits from the efforts to analyze and counter the viruses extended far beyond just the act of containment, pushing technological advances in a broader context.

Even today, examining the early history of viruses provides valuable insights into the fundamental principles of malware development and propagation. It reminds us of the constantly evolving nature of this threat landscape and the need for ongoing vigilance. Understanding the genesis of these digital organisms provides an essential context for comprehending the far more sophisticated threats we face now. We can learn from the mistakes of the past to better address the challenges of the future, and avoid repeating the same vulnerabilities. The lessons learned from those early days continue to inform modern cybersecurity strategies and practices.

2.2 The Rise of Worms

The Morris worm, unleashed in 1988, serves as a pivotal moment. Its relatively simple design, exploiting a vulnerability in the finger daemon, demonstrated the terrifying potential of self-replicating malware. Unlike viruses needing a host program to spread, the worm's autonomous propagation across the nascent internet highlighted the vulnerability of interconnected systems. Its impact was profound, crippling a significant portion of the then-existing internet infrastructure, a stark warning of the power of even rudimentary worm technology. The event underscored the need for robust network security measures, a lesson still acutely relevant today. This unforeseen disruption, a precursor to future, more sophisticated attacks, proved that network security wasn't merely a technical challenge, but a matter of national and global importance.

The conceptual leap from the Morris worm to later, more sophisticated iterations wasn't a giant bound, but rather a series of incremental improvements. Programmers, driven by an array of motivations—from simple curiosity to malicious intent—exploited newly discovered vulnerabilities, refining worm designs to enhance their stealth, speed of replication, and capacity for damage. This organic evolution, fueled by constant innovation and exploitation of software weaknesses, produced worms that were far more insidious and impactful than their predecessor. The development of these sophisticated iterations wasn't confined to a single laboratory or nation; it was a distributed, often anonymous, process characterized by the sharing of techniques and code amongst a community of diverse actors.

Understanding worm propagation requires a deeper look into network topologies and vulnerabilities. Worms, unlike viruses, don't require user interaction for infection. Their propagation methods are often based on exploiting weaknesses in network services, such as open ports or outdated software. The sheer scale of the internet, with its multitude of interconnected systems, provided a fertile ground for worm proliferation. Once a single infected machine is compromised, the worm can rapidly traverse the network, infecting countless others, and amplifying its damage exponentially. This ease of spread distinguishes worms from other malware types and emphasizes the importance of network segmentation and regular software patching. Indeed, a significant portion of worm attacks could be prevented with proactive security practices.

The impact of a successful worm attack goes far beyond simple system disruptions. Data breaches, system instability, service outages, and financial losses are commonplace consequences. The disruption of critical infrastructure, such as power grids, transportation networks, and financial institutions, presents a significant national security risk. This is not merely a theoretical concern; history is filled with instances where malicious worms have crippled businesses, governments, and entire societies. A recent example highlights the growing threat: a worm that specifically targeted industrial control systems was recently discovered, capable of causing physical damage and significantly disrupting critical processes. The potential impact of these sophisticated worms underscores the growing need for robust security measures in the era of increasingly interconnected systems.

The evolution of worms isn't just a technological phenomenon; it's also a reflection of the ever-evolving arms race between those who create malware and those who seek to defend against it. As defensive measures improve, malware creators develop increasingly sophisticated techniques to circumvent them. This ongoing cycle of adaptation necessitates a dynamic approach to cybersecurity, emphasizing continuous learning, adaptation, and collaboration between researchers, security professionals, and policymakers. This ongoing back-and-forth highlights the critical importance of staying ahead of the curve in the dynamic world of cybersecurity. The fight against advanced worms requires a constant effort towards enhancing security infrastructure and promoting a collective understanding of the ever-changing threats.

This constant evolution underlines the crucial need for a proactive, multi-layered approach to network security. It's no longer sufficient to rely solely on perimeter defenses. Effective countermeasures must involve robust internal security practices, regular software updates, employee training programs that promote strong cybersecurity awareness, and effective incident response planning. We must not only anticipate the spread of known vulnerabilities but also invest in research and development to stay ahead of emerging threats. The sophisticated nature of modern worms demands a sophisticated, layered, and adaptable defense strategy. The stakes are high, and complacency is simply not an option.

2.3 Trojans and Their Tactics

Trojan horses, unlike their mythical counterparts, don't offer gifts of beauty or power. Instead, they deceptively infiltrate systems, posing as legitimate software or files, a wolf in sheep's clothing, quietly waiting to unleash their malicious payload. Their insidious nature makes them particularly dangerous, often bypassing initial security defenses through social engineering or software vulnerabilities. The user, unknowingly, becomes the key to their success.

Understanding their tactics requires a deep dive into their methods. They might arrive disguised as a seemingly harmless email attachment, a cracked software download, or a seemingly helpful update. Once installed, the Trojan's actions can vary wildly. Some simply open backdoors for remote access, allowing attackers to control the compromised system at will. This control can lead to data theft, system manipulation, or even the installation of further malware. Others might silently log keystrokes, capturing passwords and sensitive information. Still others act as components of larger botnets, their individual actions seemingly insignificant until they're orchestrated as part of a coordinated attack. The diversity of their capabilities highlights the need for robust security practices.

The deceptive nature of Trojans necessitates a proactive approach to security. Users should cultivate a healthy skepticism towards unsolicited emails and attachments. Verifying the legitimacy of software downloads from reputable sources is paramount. Regularly updating software patches that address known vulnerabilities is vital in preventing attackers from exploiting weaknesses and introducing Trojans. These simple steps significantly reduce the risk of infection.

Moreover, understanding the various attack vectors employed by Trojans is essential for effective defense. Social engineering, the art of manipulating individuals into divulging confidential information or performing actions that compromise security, plays a significant role. Attackers may use phishing emails, fabricated online personas, or even phone calls to trick unsuspecting users into installing malware. Recognizing the hallmarks of such scams is crucial, requiring awareness of common phishing techniques and a critical eye for inconsistencies in communication.

Beyond social engineering, Trojans also exploit software vulnerabilities. Outdated software often contains security flaws that attackers can exploit to gain unauthorized access. Regular updates provide critical security patches that address known vulnerabilities, rendering many Trojan attack vectors ineffective. Staying informed about emerging threats and vulnerabilities is vital, enabling prompt updates and mitigation of potential attacks.

Furthermore, the sophisticated nature of modern Trojans demands advanced security measures. Robust antivirus software, regularly updated and capable of detecting and removing malware, is a fundamental component of a strong security posture. Network security solutions, such as firewalls and intrusion detection systems, can provide additional layers of protection by monitoring network traffic and blocking malicious activity. Regular network security assessments can identify vulnerabilities and help maintain a strong security perimeter.

Finally, effective security is not solely reliant on technology. Security awareness training plays a vital role in educating users about the threats posed by Trojans and other malware. Training programs should emphasize safe internet practices, the importance of software updates, and the risks associated with clicking on suspicious links or downloading attachments from unknown sources. Empowering users with the knowledge to identify and avoid threats is a critical aspect of building a robust security defense. The combination of technology and user awareness provides the best defense against the stealthy attacks of Trojans. Staying informed, remaining vigilant, and practicing safe computing habits are the best lines of defense against these cunning digital threats. They are not invincible, and with the right knowledge and tools, you can significantly reduce your risk.

3 Sophisticated Malware Techniques

Polymorphism and metamorphism are crucial techniques employed by sophisticated malware to evade detection. Polymorphic malware changes its code structure with each infection, rendering signature-based detection ineffective. Think of it like a chameleon adapting to its environment; the underlying functionality remains, but the outward appearance constantly shifts. Metamorphic malware, on the other hand, alters its code's functionality while maintaining its core malicious behavior. This is a more advanced approach, making analysis significantly more complex. Understanding these techniques is fundamental to developing effective countermeasures. Researchers continuously adapt their methodologies to keep pace with this ever-evolving threat.
Rootkits represent a significant advancement in malware stealth. Unlike simpler malware, rootkits actively conceal their presence from the operating system and security software. They achieve this by modifying system files, hooks in system calls, and other low-level manipulations. This makes detection exceptionally challenging, often requiring deep system analysis. The effect is insidious; a compromised system appears normal, yet a hidden process is actively working in the background, potentially exfiltrating data, providing backdoor access, or facilitating other malicious actions. Discovering and removing a rootkit requires specialized tools and

expertise.
Advanced evasion techniques are a hallmark of modern malware. These go beyond simple code obfuscation. Sophisticated malware might employ techniques like process injection, where the malicious code injects itself into legitimate processes, masking its origin. Other methods involve exploiting system vulnerabilities or employing advanced anti-debugging techniques, hindering attempts to analyze the malware's behavior. The goal is simple: to remain undetected, allowing the malware to operate unchecked within the system. This constant arms race between attackers and defenders drives continuous innovation in security technologies.
Data exfiltration methods represent the means by which malware achieves its ultimate goals, typically involving stealing sensitive data. Malware might use various techniques, from simple email attachments to more sophisticated methods like using proxy servers or encrypted communication channels. The choice of method often depends on the type of data being targeted and the level of sophistication the attacker employs. Data exfiltration can occur silently and undetected, allowing sensitive information to be transferred to remote locations for later use. Effective security measures need to focus not just on preventing infection, but also on monitoring network traffic for unusual data flows and securing the data itself.
Botnet architectures are a critical aspect of many large-scale malware campaigns. A botnet comprises many compromised systems, controlled by a centralized command-and-control server. This server allows the attacker to coordinate attacks, such as distributed denial-of-service (DDoS) attacks, or to launch large-scale spam campaigns. The architecture is designed for resilience and scalability, making it difficult to dismantle. The compromised systems, or bots, remain largely undetected, their resources effectively hijacked for malicious purposes. Identifying and disrupting the command-and-control server is crucial to neutralizing the threat, but the decentralized nature of some botnets makes this task challenging. Moreover, the sheer scale of some botnets poses significant challenges for traditional security systems.

3.1 Polymorphism and Metamorphism

Polymorphism, in the context of malware, is a technique where the malicious code alters its appearance without changing its functionality. Imagine a chameleon; it changes its color to blend in with its surroundings, but it remains the same chameleon. Similarly, polymorphic malware modifies its code's structure—the sequence of instructions, the use of encryption, or even the names of its components—on each infection, making it difficult to detect using traditional signature-based antivirus solutions. These signatures are essentially "fingerprints" of known malware; if the fingerprint changes, the antivirus misses it. The challenge lies in the sheer number of variations a polymorphic program can generate. A single piece of polymorphic malware can produce millions of unique variants, requiring continuous updating of antivirus databases, a battle many security vendors find themselves perpetually fighting. This constant evolution necessitates a robust, multi-layered approach to security rather than relying on a single, signature-based solution.

Metamorphism takes polymorphism a step further. Instead of simply altering its appearance, metamorphic malware fundamentally changes its code during each infection, rebuilding itself from scratch while preserving its core functionality. This is analogous to rebuilding a house using different materials and architectural styles, yet maintaining the same rooms and basic structure. It doesn't just shuffle the code around; it actively transforms the underlying instructions using complex algorithms, making it almost impossible to detect through simple pattern matching. The process involves self-modification routines built into the malware itself, transforming its internal structure to a point where simple comparisons against previously observed signatures are completely ineffective. This is a major step up from polymorphism, requiring far more sophisticated analytical techniques for detection and analysis.

Understanding the differences between polymorphism and metamorphism is crucial for developing effective countermeasures. Poly-

morphic malware is easier to detect compared to metamorphic malware because the core functionality, though disguised, remains consistent. Analyzing several variants might reveal underlying similarities. However, metamorphic malware presents a considerably greater challenge due to its fundamental code alteration. Therefore, a combination of techniques such as behavioral analysis—observing the malware's actions rather than just its code—sandboxing—running the malware in a controlled isolated environment—and heuristic analysis—identifying suspicious patterns and behaviors—is often necessary to effectively detect and mitigate these advanced threats.
The impact of these techniques extends beyond simple antivirus evasion. Polymorphic and metamorphic malware increases the cost and complexity of threat detection and response. Security professionals must invest in more sophisticated analysis tools, regularly update their security infrastructure, and adapt their detection strategies constantly to keep up with the ever-evolving landscape of malware techniques. The constant arms race between malware developers and cybersecurity professionals necessitates a proactive approach, with a focus on advanced threat intelligence, automated response systems, and thorough security audits. The economic consequences of a successful polymorphic or metamorphic malware attack can be devastating, ranging from data breaches and financial losses to significant disruptions in business operations.
One specific aspect to consider regarding both polymorphic and metamorphic malware is the use of encryption. Malware often employs encryption techniques to further obfuscate its code, making static analysis considerably more challenging. The decryption process frequently requires intricate reverse engineering, and even then, the resulting code may still be heavily obfuscated. The increasing sophistication of these encryption methods requires the development of more advanced decryption tools and techniques. In fact, researchers are continually exploring new ways to break through these layers of obfuscation, using methods such as dynamic analysis where the malware is run in a safe environment and its be-

havior is observed while attempting to decrypt parts of the code. However, malware developers continue to refine their encryption techniques, turning this into another ongoing battle of innovation. Therefore, a multi-pronged approach, combining static, dynamic, and behavioral analysis techniques, is the most effective way to combat these evolving threats. The evolution of encryption within malware highlights the importance of not relying on a single detection method but employing a broader, layered security strategy.

3.2 Rootkit Technology

Rootkits represent a particularly insidious class of malware, designed not just to infect a system but to remain completely hidden from detection. Unlike viruses that might trigger obvious system slowdowns or error messages, rootkits operate in the shadows, silently manipulating system processes and data. Their stealthy nature makes them incredibly dangerous, as they can persist for extended periods, enabling attackers to maintain persistent access and control. Imagine a burglar not just stealing your valuables, but also installing hidden cameras and a secret back door, allowing them to return at will without your knowledge. That's the essence of a rootkit.

The core functionality of a rootkit revolves around its ability to conceal its presence. This is achieved through a variety of techniques, from hiding files and processes in the operating system's file system and registry to modifying system calls and API hooks. Rootkits might alter the output of system monitoring tools, making them appear as though nothing is amiss. They effectively lie to the system, presenting a false reality to any attempts at investigation. Consider the analogy of a magician's sleight of hand, skillfully manipulating perceptions to mask their actions; rootkits employ a similar level of deception.

One of the most common methods employed by rootkits involves the manipulation of the operating system's kernel. By hooking or modifying core kernel functions, a rootkit can intercept and alter system calls, preventing security software from detecting its presence or activities. This deep level of access provides the rootkit with unparalleled control, enabling it to modify system settings, steal data, and execute arbitrary commands without raising suspicion. This intricate level of manipulation makes detection and removal exceptionally challenging.

Furthermore, many rootkits utilize advanced techniques to prevent their analysis. These techniques often involve code obfuscation, polymorphism, and anti-debugging measures. Code obfuscation makes the rootkit's code difficult to understand, making reverse engineering a formidable task. Polymorphism allows the rootkit to change its code signature, evading detection by antivirus software that relies on signature-based detection methods. Anti-debugging techniques actively detect and prevent debugging attempts, hindering efforts to understand the rootkit's behavior and functionality.

The impact of a rootkit infection can be devastating. Compromised systems can be used for various malicious activities, including data theft, identity theft, launching attacks against other systems, and even acting as part of a larger botnet. Because the rootkit is hidden, the attacker can maintain persistent access, potentially for months or even years, without the system owner's knowledge. This prolonged access can lead to significant data breaches, financial losses, and reputational damage.

Therefore, robust security practices are crucial in mitigating the risk of rootkit infections. This involves regularly updating operating systems and software, employing strong antivirus and anti-malware solutions, and implementing robust security protocols. Regular security audits and penetration testing can help identify vulnerabilities that might be exploited by rootkits. Educating users about potential threats and promoting safe computing practices is also essential. Ultimately, a layered security approach, combining multiple defensive measures, offers the best protection against these elusive threats. Remember, prevention is far easier, and significantly less costly, than remediation after a successful rootkit

infection. The potential consequences are simply too high to ignore.
Beyond these basic preventative measures, proactive monitoring is essential. Regularly checking system logs for unusual activity, scrutinizing network traffic for suspicious patterns, and employing tools specifically designed to detect rootkits can significantly improve security posture. Understanding the typical behaviors associated with rootkit infections, such as unusual system performance, unexpected network activity, or the appearance of unknown processes, can provide early warning signs. The sooner a rootkit is detected, the easier it is to contain its impact and prevent further compromise.

3.3 Advanced Evasion Techniques

Obfuscation techniques go beyond simple encryption; they involve manipulating the malware's code to make it virtually unrecognizable to antivirus software. This might include using code packers to compress and encrypt the executable, making reverse engineering extremely difficult and time-consuming. Furthermore, sophisticated malware often employs polymorphic and metamorphic engines, constantly changing its code while maintaining its functionality. This dynamic nature frustrates signature-based detection, rendering traditional antivirus solutions ineffective. The sheer volume of variations generated, along with their rapid mutation rate, overwhelms static analysis methods.
Another potent evasion tactic hinges on mimicking legitimate system processes. Malware writers craft their creations to masquerade as essential system files or applications, thus blending seamlessly into the background. This deception is further enhanced by the use of rootkits, which hide the malware's presence entirely from the operating system and other security tools. Rootkits often modify the system's API calls, preventing the detection of malicious processes or files. They're difficult to detect and even more challenging to

remove, requiring specialized tools and expertise. The clandestine nature of rootkits, combined with their ability to manipulate the very core of the operating system, makes them incredibly effective at achieving persistent infection.

Advanced evasion techniques often involve exploiting system vulnerabilities and manipulating operating system internals. This includes using undocumented functions or leveraging flaws in the kernel to bypass security checks and restrictions. Malware might modify the system's boot process, ensuring its persistence even after a system reboot. It might even interact directly with hardware components, potentially leaving no trace in the software environment. Such techniques require a deep understanding of the operating system's architecture and its potential weaknesses, placing them beyond the capabilities of amateur malware creators. The level of skill required for this level of manipulation underscores the constant arms race between attackers and defenders.

Network-based evasion involves various techniques to conceal malicious communications. Malware might use encrypted channels to transmit data, preventing network security tools from inspecting its traffic. It could also utilize proxy servers or VPNs to mask its origin and destination, making it practically untraceable. Furthermore, malware can leverage encrypted tunneling protocols to bypass firewalls and intrusion detection systems, creating covert channels for data exfiltration. This creates a significant challenge for organizations attempting to monitor their network traffic for suspicious activities, demanding sophisticated network monitoring solutions capable of detecting hidden communications. The complexity of such networks requires a multifaceted approach to security.

The ability to avoid sandboxing is crucial for advanced malware. Sandboxing techniques are commonly used in malware analysis to isolate malicious code and observe its behavior in a controlled environment. However, sophisticated malware can detect when it's running inside a sandbox, and alter its behavior accordingly. This might involve checking for the presence of debugging tools or specific API calls characteristic of sandboxed environments. If the

malware determines it's being analyzed, it might simply remain dormant or cease its malicious activities, thus evading detection. This adaptability renders simple sandbox analysis often insufficient, requiring more complex and dynamic analysis approaches.
Furthermore, malware might use code virtualization techniques to evade detection. By running its code within a virtual machine, the malware effectively isolates itself from the host system. This prevents direct access to the malware's code, making reverse engineering extremely difficult. This approach adds another layer of complexity to the already challenging process of malware analysis, requiring specialized tools and deep expertise to overcome. The resilience and adaptability of these techniques showcase the sophistication and resourcefulness of modern malware creators.
Finally, the use of anti-analysis techniques actively hinders reverse engineering efforts. Malware might incorporate self-modifying code, making static analysis incredibly difficult. It might also use code obfuscation techniques to scramble the code, making it harder to understand its functionality. The use of packers further complicates the process, requiring unpacking before any meaningful analysis can occur. Anti-debugging techniques further complicate matters, thwarting attempts to attach debuggers to the running process, thus making it difficult for security researchers to investigate malware's behavior. This constant evolution in evasion techniques underscores the ongoing need for advanced malware analysis tools and expertise. The constant arms race between attackers and defenders underscores the continuous need for adaptation and innovation.

3.4 Data Exfiltration Methods

Data exfiltration, the clandestine removal of sensitive data from a compromised system, is a crucial stage in many sophisticated malware attacks. Imagine a thief not just breaking into a house, but meticulously packing valuables and leaving undetected. That's essentially what data exfiltration represents for cybercriminals. Their methods are constantly evolving, requiring a thorough understanding to effectively counter them.

One common technique leverages legitimate channels. Malware might use compromised email accounts to send stolen data in seemingly innocuous attachments or emails. It could mimic normal business processes, masking malicious activity within routine data transfers. This method relies on the attacker's ability to blend into the background noise of legitimate network traffic, making detection extremely challenging. The success depends on the attacker's meticulous planning and execution, mimicking regular network activity with deceptive accuracy.

Alternatively, malware can use covert channels. These channels operate outside standard communication pathways, hiding data within seemingly harmless network traffic or using unconventional methods like exploiting vulnerabilities in system protocols to smuggle information. Think of it as creating secret passages within the network, bypassing typical security checkpoints. This requires significant technical expertise, utilizing vulnerabilities and exploiting their inherent weaknesses, thus requiring a profound understanding of network infrastructure.

Another approach involves using compromised servers as staging points. Stolen data is transferred to these hidden servers, acting as intermediate points before being sent to a final destination, possibly overseas or using anonymizing services, to further obfuscate the origin and destination. This multi-layered approach makes tracing the data back to its source extraordinarily difficult, even with advanced investigative tools. The attacker adds additional security by utilizing multiple servers, adding an extra level of difficulty for investigators.

Direct exfiltration might seem straightforward, but it often involves sophisticated techniques. This direct method involves the malware directly communicating with a command-and-control (C&C) server to upload stolen data. However, the communication itself is usually encrypted and obfuscated, employing techniques like tunneling, encryption, and steganography, making it practically invisible to casual observation. The encryption and the techniques used to hide the data must be robust to withstand various attempts at interception, therefore it presents a challenging target for defensive mechanisms.

File Transfer Protocol (FTP) and HTTP are frequently exploited for exfiltration. While these protocols are legitimate, malware can abuse them by using hidden commands or manipulating requests to transfer stolen data undetected. The attacker carefully crafts the requests to look like ordinary traffic, hence making it practically indistinguishable from legitimate communications. This involves careful manipulation of existing protocols and an intimate understanding of their underlying functionalities.

The use of social media platforms and cloud storage services for data exfiltration is a rapidly growing trend. Malware may upload stolen data to seemingly innocuous accounts or cloud storage services, taking advantage of the vast amount of data already being transferred on these platforms. Detection is hampered by the sheer volume of data and the difficulty in distinguishing malicious uploads from legitimate user activity. Criminals exploit the unsuspecting nature of these services, ensuring the data transfer remains discreet.

Finally, remember that exfiltration methods are often combined. A sophisticated attack might employ multiple techniques simultaneously, layering security mechanisms to make detection and analysis exceedingly difficult. This multi-layered approach increases the resilience of the attack against detection and analysis attempts, hence the necessity of advanced techniques for security professionals to counter them effectively. Understanding these techniques is crucial for developing robust security measures and for successfully investigating sophisticated malware attacks. The evolution of data exfiltration methods requires a constant vigilance and adaptation from security professionals and researchers alike. Staying abreast of the newest techniques is an ongoing battle, crucial for maintaining a strong and secure online environment.

3.5 Botnet Architectures

Botnets represent a significant threat in the modern cybersecurity landscape, acting as distributed networks of compromised computers controlled by a single attacker, or Command and Control (C&C) server. Their architecture is crucial to understanding their capabilities and resilience. Understanding their structure is key to developing effective countermeasures.

The simplest botnet architecture utilizes a centralized C&C server. This server directly communicates with each infected machine, or bot, issuing commands and receiving data. While straightforward, this model is vulnerable to takedowns; removing the central server effectively disables the entire botnet. However, the simplicity makes it a starting point for many aspiring attackers. This architecture often uses simple protocols such as IRC, making them comparatively easier to identify and analyze.

A more resilient design involves a hierarchical structure, incorporating multiple layers of C&C servers. This decentralized model distributes command dissemination, increasing the botnet's robustness against takedown attempts. Disrupting one layer might only partially affect the botnet's operations. The communication between layers can be sophisticated, potentially utilizing encryption and tunneling techniques to evade detection. Furthermore, the hierarchical nature makes identifying the root of the network very difficult; the main attacker may be layers away from the directly controlled bots, making it harder for law enforcement to trace and prosecute.

Peer-to-peer (P2P) botnets represent a significant advancement in resilience. In this architecture, bots communicate directly with each other, eliminating the single point of failure inherent in centralized and hierarchical models. This decentralized nature significantly increases the botnet's survivability, making it far more challenging to disable. A P2P botnet can be thought of as a constantly shifting and adapting network; it is highly dynamic, making it challenging to monitor and detect. The communication protocols used are often more sophisticated and less obvious than those in centralized structures, leveraging advanced encryption techniques to hide malicious activity.

Hybrid botnet architectures combine elements of centralized, hierarchical, and P2P models, offering a level of complexity and robustness that poses a formidable challenge. These sophisticated models usually involve a combination of the previous architectures, adapting dynamically to circumstances. They might use a centralized server for overall coordination while employing P2P for distributing specific tasks or communicating among subsets of bots. This hybrid model offers the best of multiple architectures—the control of a centralized model with the resilience of a P2P structure—making it very difficult to combat effectively.

The choice of architecture significantly impacts a botnet's capabilities. Centralized models may be simpler to manage but are more vulnerable to disruption. Hierarchical structures offer a balance between control and resilience, while P2P architectures prioritize resilience and anonymity but require more sophisticated management. Hybrid architectures, the most sophisticated approach, offer a combination of all strengths but require the most effort and expertise to build and operate.

Understanding these architectural differences is critical for both security professionals and researchers. Effective countermeasures must adapt to the specific architecture of a botnet. For instance, disrupting a centralized botnet requires targeting the central server, while combating a P2P botnet necessitates identifying and isolating key nodes within the network. Sophisticated analysis techniques, including network traffic analysis and malware reverse engineering, are crucial for identifying and characterizing the architecture of a given botnet. Future research should focus on developing techniques for identifying and mitigating the growing sophistication of hybrid botnet architectures. This is especially important given their capacity to adapt and change their structure and operations dynamically. It is clear that the architecture of a botnet significantly dictates its resilience and capabilities, making a comprehensive understanding of these architectural choices critical in developing effective defensive strategies.

4 Malware Analysis Methods

Understanding a malware sample requires a multifaceted approach, often combining several techniques. Static analysis, the examination of a malware sample without executing it, offers a first glimpse into its structure and potential capabilities. We can use tools like disassemblers to scrutinize the code, identifying functions, strings, and potentially malicious code sections. This provides a foundational understanding before proceeding to more involved methods. Remember, static analysis alone cannot reveal all the intricacies of a sample's behavior; it's merely a starting point. Careful consideration of file signatures, packers, and obfuscation techniques is crucial during this stage. Misinterpreting packed code, for instance, can lead to flawed conclusions. Therefore, meticulous attention to detail is paramount.

Dynamic analysis involves executing the malware sample in a controlled, isolated environment, such as a virtual machine or sandbox. This allows researchers to observe the malware's behavior in real-time, identifying its actions, network connections, and registry modifications. Tools like debuggers and system monitors are invaluable here, providing a detailed trace of the malware's activity. Careful monitoring of system calls and network traffic helps reveal the malware's ultimate objectives. This approach is particularly helpful in understanding the malware's actions post-infection, giving investigators crucial insights into data exfiltration methods or persistence mechanisms. Analyzing logs for suspicious activities, like unauthorized access attempts or unusual system changes, is also important, aiding in understanding the full impact of the malware.

The choice between static and dynamic analysis often depends on the specific malware sample and the analyst's goals. Some samples may be heavily obfuscated, making static analysis extremely challenging. Others may require dynamic analysis to observe their full behavior, particularly those that employ advanced anti-analysis techniques. Ideally, a combination of both techniques provides the most complete understanding. A strong understanding of assembly language is fundamental for in-depth analysis, allowing for a more comprehensive interpretation of the malware's code, especially when dealing with heavily obfuscated samples.

However, even with the most thorough analysis, some malware may remain elusive, employing advanced techniques to thwart detection. This necessitates staying updated with the latest malware trends and analysis techniques. Continuous learning and the adoption of new tools are key to effectively countering the ever-evolving landscape of malware. The effectiveness of any analysis is further enhanced through collaboration and sharing of information amongst experts in the cybersecurity community, leading to faster identification and mitigation strategies. Developing and maintaining a robust malware analysis laboratory is crucial; the lab should be equipped with up-to-date tools and resources for static and dynamic analysis, as well as a network environment for mimicking real-world scenarios. Furthermore, ethical considerations should be taken into account while dealing with malicious software, ensuring responsible and legal handling of the samples. The security and integrity of the analysis environment are crucial, preventing the analyzed malware from escaping containment and compromising the entire system. Regularly updating and patching the analysis environment is essential for minimizing any such risks.

Ultimately, mastering malware analysis demands patience, persistence, and a keen eye for detail. It is a continuous process of learning and adaptation, demanding both technical expertise and a systematic approach. Understanding the strengths and limitations of various techniques, integrating those techniques correctly, and regularly updating your skillset are crucial to remaining effective against the constantly evolving threats posed by sophisticated malware. While the techniques described above are indispensable, effective countermeasures also involve robust network security protocols, comprehensive endpoint protection, and security awareness training, thus creating a layered approach to combat the many challenges posed by evolving malware. This approach minimizes the risk of infection and maximizes the efficacy of response actions when an incident does occur. Regularly reviewing security logs and conducting penetration testing are also crucial aspects of a robust defense strategy, enabling proactive identification of vulnerabilities and strengthening network defenses.

4.1 Static Analysis

Static analysis is a crucial first step in understanding malware. It's like carefully examining a suspect's photograph before questioning them – you gather essential information before potentially unleashing something dangerous. Unlike dynamic analysis, which involves running the malware, static analysis dissects the code without executing it. This non-invasive approach minimizes risks, making it safer for researchers and analysts. We're essentially peering into the malware's blueprint, identifying its components and potential behavior before it even has a chance to wreak havoc.

This process often begins with unpacking the malware, a procedure that involves removing any layers of obfuscation or encryption the malware might employ. Think of it as carefully peeling back the layers of an onion, revealing its core structure. This can be surprisingly complex, with some malware using multiple layers to mask its true intentions and capabilities. Once unpacked, analysts can use various tools and techniques to investigate the code's structure. Disassemblers translate machine code into human-readable assembly language; debuggers allow the stepping through of code in a controlled setting; and hex editors give a low-level view of the file's contents. Each tool provides valuable insights into different aspects of the malware, and experienced analysts will often use several in concert to create a holistic understanding.

String analysis, a significant part of static analysis, involves searching for textual information within the malware's code. This can reveal critical information, such as command-and-control server addresses, API calls used by the malware, or even strings revealing the malware's purpose or origin. These seemingly small clues can be vital pieces of the puzzle, guiding the investigation and helping researchers paint a complete picture of the malware's functionality and intentions. Imagine discovering a hidden message within a seemingly innocuous document – a similar sense of discovery is often experienced while performing string analysis.

Beyond simple text searches, analysts use more sophisticated techniques. Control-flow analysis examines the execution path of the malware's code, mapping out the flow of logic. This process can reveal the conditions under which certain functions are executed, providing insights into the malware's decision-making process. It's like charting the course of a ship; understanding the navigational instructions helps us anticipate its destination. Data-flow analysis, on the other hand, tracks how data is passed between different parts of the code. This analysis identifies how data is manipulated and used by the malware, helping us understand what actions the malware will take on a system. It's the crucial step in mapping the data streams.

Furthermore, API calls made by the malware can provide valuable clues about its capabilities and behavior. These API calls show how the malware interacts with the operating system and applications, and can reveal whether it attempts to access sensitive data, modify system settings, or communicate with external servers. Analyzing this interaction can help analysts anticipate the malware's next moves and build a comprehensive profile. This insight is critical for understanding its potential destructive capacity and the level of risk it poses. Identifying the malware family through signature-based detection, by comparing the code to known malware signatures in databases, is another crucial technique. This often provides a quick identification and helps prioritize further analysis.

Finally, understanding the overall architecture of the malware and identifying its various components is paramount. This often requires a synthesis of the various analysis techniques discussed above, bringing together the information gathered from disassembling, string analysis, and control and data flow analysis. By piecing together the different puzzle pieces, analysts can build a clear picture of the malware's behavior, its purpose and its potential impact. This holistic approach is essential for developing effective countermeasures and preventing future attacks. With the help of these static techniques, we are now armed to deal with some of the insidious threats plaguing the digital world, ensuring a safer future for all.

4.2 Dynamic Analysis

Dynamic analysis offers a powerful, albeit resource-intensive, approach to understanding malware behavior. Unlike static analysis, which examines code without execution, dynamic analysis involves running the malware sample in a controlled environment, typically a sandbox, to observe its actions firsthand. This allows for the identification of behaviors that are not readily apparent from the code itself, such as network connections, file system modifications, and registry changes. Careful observation unveils the malware's true capabilities and intentions, providing crucial insights for effective mitigation and prevention.

The process begins with selecting an appropriate sandbox environment. This crucial step minimizes the risk of the malware affecting your system. Virtual machines (VMs) provide a robust, isolated space, allowing researchers to observe malware activity without fear of compromising their own systems. Sophisticated sandboxes even offer network simulation capabilities, enabling researchers to mimic real-world network conditions and further observe the malware's interactions. The choice of sandbox should align with the suspected complexity and potential destructive capability of the malware.

Analyzing the malware's network activity is paramount. Modern malware often communicates with command-and-control (C2) servers, exfiltrating stolen data or receiving further instructions. Monitoring network traffic within the sandbox reveals the malware's communication patterns, including IP addresses, ports, and protocols used. This information identifies the C2 servers and helps understand the malware's overall goals, aiding in identifying potential victims and disrupting its operations. Careful examination of network packets allows for decryption of communication if necessary.

System-level changes are another key focus. Many malware samples modify the system's registry, create new files and directories, and even install themselves as system services. Monitoring these changes provides an excellent indicator of the malware's persistence mechanisms. This detailed examination reveals how the malware attempts to maintain its presence on the compromised system, highlighting crucial weaknesses that can be exploited for removal. Regular snapshots of the sandboxed system aid in this meticulous process of detection.

Memory analysis adds a critical dimension to understanding malware behavior. Malware often uses sophisticated techniques to hide its presence in memory, making detection difficult. Advanced memory forensics tools, combined with knowledge of common malware techniques, provide the necessary tools to uncover these hidden activities. These techniques can identify malicious code execution, detect obfuscation attempts, and reveal the malware's true nature even when it's actively attempting to conceal itself. This layer of analysis frequently uncovers advanced evasion tactics that static

analysis often misses.
Process monitoring offers a dynamic perspective on malware activity. By tracking the malware's processes, their parent processes, and their children, researchers can build a clear picture of how the malware interacts with the system. This includes identifying the processes it uses, the libraries it calls, and the files it accesses. Tracking these activities reveals the malware's functions and allows for a comprehensive understanding of its lifecycle. Using specialized process monitoring tools in conjunction with memory analysis provides a very detailed view of the malware's execution path.
Finally, the use of debuggers allows for step-by-step analysis of the malware's code execution. This technique, while more complex and requiring more expertise, offers an exceptionally high level of granularity in understanding the intricacies of malware function. By meticulously stepping through the code, analysts gain insights into the underlying logic, uncovering how various components interact and achieve their malicious goals. This approach unveils sophisticated evasion tactics and provides invaluable data for crafting more effective countermeasures. However, it requires a significantly higher level of skill and expertise. This granular level of insight is critical for understanding advanced malware that use sophisticated obfuscation and anti-analysis techniques. The insights gained through dynamic analysis ultimately are crucial for developing effective security solutions and for understanding the complex threat landscape.

5 Network Security Measures

Implementing robust network security is crucial in the face of ever-evolving malware threats. A layered approach, combining multiple security measures, offers the strongest defense. This is not simply about installing software; it's about understanding the vulnerabilities and actively mitigating them. Neglecting even a single layer can leave your network exposed to exploitation.

Firewalls act as the first line of defense, diligently screening incoming and outgoing network traffic. They examine data packets based on predefined rules, blocking malicious activity while allowing legitimate traffic to pass. Intrusion detection systems (IDS), working in tandem with firewalls, actively monitor network traffic for suspicious patterns, providing alerts when potentially harmful activities are detected. Real-time analysis and quick response are vital for effective mitigation, potentially preventing significant damage before it occurs. Remember, regular updates to both firewalls and IDS are paramount; manufacturers constantly release patches to address newly discovered vulnerabilities. Failing to update exposes your systems to known exploits.

Endpoint security goes beyond network perimeters. It focuses on securing individual devices, such as workstations, servers, and laptops, which are often the entry points for malware. Employing robust antivirus software, regularly updated with the latest virus definitions, is a fundamental aspect of endpoint security. This software acts as a final line of defense against malicious code that has bypassed network security measures. Implementing strong access controls, including multi-factor authentication, further enhances security. This adds multiple layers of verification, significantly reducing the chances of unauthorized access. Regular software updates and patching across all devices should never be overlooked, addressing vulnerabilities before malicious actors can exploit them. Consider employing endpoint detection and response (EDR) solutions; these offer advanced threat detection and response capabilities.

Security awareness training is often the most underestimated yet arguably the most critical security measure. It empowers users to be the first line of defense. Training programs should focus on educating users about phishing scams, malware distribution methods, and safe browsing practices. Emphasizing the importance of strong passwords, avoiding suspicious links, and reporting suspicious activity are crucial components of effective security awareness training. Regular training sessions, perhaps incorporated into ongoing professional development, should reinforce these concepts. The training must be practical and relevant, reflecting the types of threats your specific organization faces. Consider using realistic simulations, such as mock phishing emails, to help users recognize and avoid these common attacks. Remember, a well-trained workforce is far less likely to fall victim to social engineering tactics. The costs associated with a data breach far outweigh the investment in such training.

Building a resilient network demands a multifaceted strategy. It's not a one-size-fits-all solution; your approach must align with the specific risks your organization faces. Regularly assessing your vulnerabilities and adapting your security measures accordingly is crucial for long-term protection. This is an ongoing process, requiring continuous monitoring, evaluation, and adaptation in response to the ever-changing threat landscape. This continuous adaptation, more than anything, will help your network withstand the most sophisticated attacks. Remember, complacency is a cybersecurity weakness. By prioritizing proactive measures and maintaining a vigilant approach, you can significantly reduce the risk of a devastating malware attack. Don't let your network become the next headline.

5.1 Firewalls and Intrusion Detection

Firewalls act as the first line of defense, meticulously scrutinizing all network traffic entering and leaving your system. Think of them as highly trained security guards, diligently checking every individual's identification before granting access to a secure building. They examine data packets, comparing them against predefined rules to determine whether they're safe or pose a threat. These rules, configurable by system administrators, define what type of traffic is permitted or blocked. A well-configured firewall can effectively prevent many common attacks, such as unauthorized access attempts and malicious code injections. However, even the most sophisticated firewall isn't impenetrable; it relies on correctly defined rules and regular updates to maintain its effectiveness. Failing to keep the firewall rules current leaves gaping security holes, potentially exposing your network to advanced threats that exploit zero-day vulnerabilities, bypassing the existing protective measures. Neglecting this crucial aspect of network security can have devastating consequences, resulting in data breaches or system compromises, ultimately crippling your organization's operations. Regular updates are therefore not optional, but essential for the continued well-being of your digital infrastructure.
Intrusion Detection Systems (IDS), unlike firewalls that prevent malicious traffic, primarily focus on detecting malicious activity already present within the network. They work more like surveillance cameras, monitoring network traffic and analyzing patterns for suspicious behavior. An IDS can identify anomalies, such as unusual login attempts, excessive data transfers, or attempts to ac-

cess unauthorized resources. Once detected, an alert is triggered, enabling security personnel to respond swiftly and mitigate the threat. There are two primary types: network-based IDS (NIDS), monitoring network traffic for suspicious patterns, and host-based IDS (HIDS), monitoring activity on individual systems. NIDS offer a broader view of network activity, detecting attacks that might evade individual host-based systems, while HIDS provide deeper insights into the behavior of specific machines. Combining both approaches offers comprehensive network protection, providing layered security to increase resilience against sophisticated attacks. Think of them as an advanced security system for your network, working to detect and alert you of potential threats before they can cause irreparable harm.

The effectiveness of both firewalls and IDSs depends heavily on accurate configuration and ongoing maintenance. This includes regular updates to address newly discovered vulnerabilities and evolving attack methods. For example, a firewall rule that blocks traffic from a specific IP address might become ineffective if attackers utilize a different IP address to circumvent the restriction, highlighting the importance of dynamic security strategies that can adapt to emerging threats. Similarly, an IDS needs to be updated regularly to detect and respond effectively to new attack signatures and patterns. This requires ongoing analysis of current threat landscapes, a proactive response to identified threats and continuous improvements to security policies based on emerging threats and attack vectors. Failure to keep these systems updated creates significant security vulnerabilities, making your network an easy target for attackers. Think of it as leaving your house unlocked, with all your valuable items exposed. In short, proactive management and regular maintenance are absolutely crucial for their continued effectiveness.

Implementing and managing firewalls and IDSs effectively requires specialized knowledge and expertise. Misconfigurations can render these systems ineffective or even create security loopholes, making your system more vulnerable than before. For example, overly permissive firewall rules can leave your network exposed, while overly restrictive rules can disrupt legitimate traffic and hinder productivity. It's a delicate balance that requires careful consideration and planning. Professional security consultants can provide invaluable assistance in designing and deploying these systems correctly, ensuring that your network is adequately protected. Furthermore, training your IT personnel on how to effectively manage and maintain these systems is a worthwhile investment that pays off long-term. Remember, the goal isn't just to install these technologies, but to master them, transforming them into your first line of defense against sophisticated and evolving threats. Investing in proper training and ongoing professional support is crucial to the success of your security strategy. The security of your digital infrastructure relies on having the right expertise at the helm.

5.2 Endpoint Security

Endpoint security focuses on protecting individual computers and devices from malware. This is crucial because even the strongest network defenses are useless if a single compromised machine becomes a gateway for an attacker. Think of it as the last line of defense, safeguarding the individual soldier against attack while the network security acts as the fortress walls. Effective endpoint security requires a multi-layered approach, combining various technologies and strategies.

Antivirus and anti-malware software form the first layer. These programs scan files and processes for known threats, preventing malicious code from executing. Regular updates are critical, as new malware is constantly emerging, requiring constant adaptation in the software's defenses. Failing to update exposes your systems to known vulnerabilities, making them easy targets for even unsophisticated attacks. Consider it a constantly evolving arms race; you must continually upgrade to stay ahead of the ever-changing threats.

Next comes intrusion detection and prevention systems (IDPS) specifically designed for endpoints. These systems monitor system activity, looking for suspicious behavior indicative of malware infection. Unlike simple antivirus, they analyze system actions in real-time, alerting you to potential threats even before they fully manifest. This proactive approach allows for immediate responses, minimizing potential damage. Think of it as having a guard patrolling your system, constantly vigilant for any unusual activity.

Data loss prevention (DLP) tools are essential for preventing sensitive information from leaving your devices. This becomes even more crucial when dealing with devices holding confidential data, requiring stringent security measures to prevent breaches. These solutions monitor data transfers, blocking attempts to send sensitive information outside of approved channels. This layer protects against data exfiltration, a common tactic used by sophisticated malware. It's like having a secure vault within the device, preventing unauthorized access to valuable assets.

Regular software patching and updates are not just a matter of convenience; they're a fundamental aspect of endpoint security. Software vulnerabilities are constantly exploited by attackers, making timely updates critical. This ensures that known weaknesses are addressed, reducing the attack surface and mitigating the risk of infection. It's a proactive step in hardening security by continually updating your defenses, patching the holes before they can be exploited.

User education and awareness are surprisingly powerful components of a strong endpoint security strategy. Employees often unintentionally expose their systems to malware through phishing emails or unsafe browsing habits. Regular training programs can significantly reduce human error, preventing many attacks before they even start. This emphasizes the human element in the equation; a well-trained workforce is the best defense against social engineering and other similar attacks.

Finally, consider the use of sandboxing technology. This allows for the safe execution of potentially malicious files in a virtual environment, observing their behavior without risking a full-blown infection. Analyzing potentially dangerous files in a controlled environment allows you to identify and analyze malware without causing damage to your primary system. It's like having a testing ground to analyze suspicious material before unleashing it upon your valuable data. Remember, a multi-layered approach to endpoint security is vital. No single solution is foolproof, and layering defense mechanisms drastically reduces the overall risk. Each layer acts as a fail-safe, enhancing overall security and safeguarding against various attacks. Regular reviews and adjustments of your security strategy are essential to remain ahead of evolving threats. This dynamic approach ensures that your systems are always prepared for the next wave of cyberattacks.

5.3 Security Awareness Training

Security awareness training isn't just another box to tick; it's the cornerstone of a robust cybersecurity defense. A sophisticated firewall and the most advanced endpoint protection are useless if employees click malicious links or fall prey to phishing scams. Consider this: the most secure system can be compromised by a single, unwitting click. Therefore, investing in comprehensive training isn't an expense, it's an investment in minimizing vulnerabilities.
The effectiveness of training relies heavily on its design and delivery. Forget tedious, hour-long lectures; instead, focus on interactive sessions, engaging simulations, and real-world examples. People retain information far better when actively involved, rather than passively absorbing a barrage of facts. Short, frequent training modules, woven into daily workflows, are more likely to be well received and remembered than an annual marathon. Think bite-sized learning, focused on specific threats, rather than overwhelming users with broad strokes.
For instance, you could create a series of short videos demonstrating common phishing techniques. One video might show a realistic

example of a phishing email, highlighting telltale signs like suspicious sender addresses, grammatical errors, or urgent requests for personal information. Another might focus on identifying malicious websites, explaining how to check URLs and look for security indicators such as HTTPS. These short, focused learning modules, delivered over several weeks, will prove far more impactful than a single, lengthy session.

Remember, tailoring the training to your specific workforce is crucial. A training program for a tech-savvy team will differ significantly from one designed for less tech-proficient employees. The content, approach, and delivery methods should all be carefully considered to address the knowledge base and comfort levels of each group. Avoid using jargon or complex terminology; prioritize simple, clear, and straightforward language. The goal is understanding, not intimidation.

Furthermore, regular reinforcement is key. Once the initial training is completed, the learning should continue. This could involve regular email updates with security tips, short quizzes to test comprehension, or even simulated phishing campaigns to assess the ongoing effectiveness of the training. Such measures help keep security awareness at the forefront of employees' minds and ensure that best practices are consistently applied.

Don't underestimate the power of gamification in making training engaging. Introduce friendly competitions, award points for completing modules or identifying threats, and offer small rewards for participation. This approach transforms learning into a fun and interactive experience, increasing employee engagement and knowledge retention. The use of interactive scenarios and simulations helps users apply their knowledge in a risk-free environment, building confidence and reinforcing key security principles.

Effective security awareness training requires an ongoing commitment. It's not a one-time event; it's an iterative process of education, reinforcement, and evaluation. Regular feedback from employees can reveal areas needing improvement, ensuring the training remains relevant, engaging, and ultimately effective in reducing the risk of successful cyberattacks. By making security awareness an integral part of company culture, you significantly bolster your organization's overall resilience against cyber threats. Ultimately, a well-trained workforce serves as the first line of defense, significantly reducing your organization's overall vulnerability.

Finally, remember to measure the effectiveness of your training. Track employee participation rates, quiz scores, and most importantly, the number of reported security incidents. A decrease in reported phishing attempts or malware infections serves as a strong indicator of training success. Continuously refine your program based on these metrics, ensuring that the training remains relevant, current, and aligned with evolving cyber threats. A proactive and adaptable security awareness program is essential in protecting your organization's valuable assets from the ever-evolving landscape of cyberattacks. This constant evolution ensures a consistent and effective defense against the growing sophistication of malicious actors.

6 Case Studies of Major Attacks

The Stuxnet worm, revealed in 2010, stands as a landmark in targeted malware. Its sophisticated design, utilizing zero-day exploits and a multi-stage infection process, allowed it to infiltrate Iranian nuclear facilities with devastating precision. Stuxnet's ability to subtly manipulate industrial control systems, causing centrifuges to malfunction without obvious signs of compromise, showcased a new level of malicious capability. This attack highlighted the vulnerability of critical infrastructure to highly targeted, state-sponsored attacks, forcing a reassessment of industrial cybersecurity protocols globally and prompting a significant increase in research into similar threats. The meticulous planning and execution involved underscore the potential for significant damage from highly advanced malware. Its impact extended beyond the immediate target, serving as a stark warning to organizations worldwide about the risks associated with inadequate network security and the potential for advanced persistent threats to remain undetected for extended periods.

NotPetya, unleashed in 2017, wasn't your typical ransomware. Disguised as a tax-related software update, this destructive malware quickly spread globally, crippling businesses and causing billions of dollars in damages. Unlike traditional ransomware that primarily focuses on encryption and ransom demands, NotPetya's core function was data destruction. Its self-propagating mechanism, combined with its ability to overwrite master boot records, led to widespread system failures and data loss. The attack significantly impacted numerous industries, from shipping and logistics to finance and healthcare. NotPetya's unprecedented scale and indiscriminate destruction highlighted the potential for supply chain attacks and the devastating consequences of widespread ransomware deployments. It served as a stark reminder of the interconnectedness of global systems and the need for robust disaster recovery plans. Its impact underscored the importance of regularly backing up critical data and employing stringent security measures across the entire supply chain.

WannaCry, erupting in 2017, exploited a vulnerability in older versions of Microsoft Windows. Its rapid global spread, leveraging the EternalBlue exploit leaked from the NSA, showcased the catastrophic potential of easily exploitable vulnerabilities. WannaCry's impact was felt worldwide, affecting hospitals, businesses, and critical infrastructure. The ransomware encrypted files and demanded a ransom for their release, disrupting operations and causing significant financial losses. The sheer scale of this attack emphasized the importance of timely patching and the dangers of failing to update software promptly. The incident further highlighted the global threat of ransomware and the necessity for robust cybersecurity measures across all sectors, not just those traditionally considered high-risk. The quick response by various security researchers in developing a "kill switch" that stopped the malware's spread underscored the importance of international collaboration in mitigating widespread cyberattacks.

The SolarWinds supply chain attack, discovered in 2020, represented a new sophistication in malware deployment. Compromising the SolarWinds Orion software update mechanism, attackers successfully infiltrated numerous high-profile organizations. This attack highlighted the vulnerability of software supply chains and the potential for attackers to gain access to numerous systems through a single compromised component. The scale and precision of the attack, coupled with the extended period of undetected intrusion, underscore the challenges faced in identifying and mitigating supply chain compromises. The incident prompted a significant review of software supply chain security practices and emphasized the need for stronger verification and auditing processes. The extensive impact on national security and critical infrastructure served as a clear indication of the serious threat posed by supply chain attacks and the imperative for improved security throughout the entire software development lifecycle.

The Colonial Pipeline ransomware attack in 2021 shut down a major fuel pipeline in the United States, causing widespread fuel shortages and economic disruption. The DarkSide ransomware gang successfully infiltrated the pipeline's systems, encrypting critical data and demanding a significant ransom. The incident showcased the potential for ransomware to target critical infrastructure, causing widespread disruption and impacting everyday life. The immediate impact and lasting economic consequences of the attack underscored the vulnerability of critical infrastructure to cyberattacks. It served as a potent reminder of the need to protect these systems with robust security measures, including advanced threat detection, incident response capabilities, and secure backup and recovery mechanisms. The significant ransom payment also fueled the debate surrounding the ethical implications of paying ransoms to cybercriminals. The attack further highlighted the interdependencies within our modern infrastructure and the systemic risk posed by malicious cyber actors.

6.1 Stuxnet Analysis

Stuxnet, discovered in 2010, wasn't just another worm; it represented a significant leap in the sophistication of cyber warfare. Its primary target was Iran's nuclear enrichment program, specifically the Natanz uranium enrichment facility. Unlike previous malware, Stuxnet showcased a level of precision and stealth previously unseen. Its creators meticulously crafted a weapon designed to subtly sabotage industrial centrifuges, causing them to malfunction without triggering obvious alarms. This wasn't a brute-force attack; it was a surgical strike aimed at a specific, critical infrastructure component.

The malware's complexity was breathtaking. It employed multiple zero-day exploits—vulnerabilities unknown to the affected systems' manufacturers—allowing it to bypass standard security measures. This intricate approach highlights a significant shift in malware development, moving beyond simple infection and data theft towards targeted destruction of physical infrastructure. Its multi-stage infection process, coupled with its sophisticated evasion techniques, demonstrated a keen understanding of industrial control systems (ICS) and their vulnerabilities. Its ability to modify the control system's functionality through the manipulation of low-level system functions indicated a level of deep system knowledge normally associated with highly skilled engineers or insiders.

Stuxnet's propagation relied on a combination of removable drives and network-based infection. This method, often described as a dual vector approach, increased the likelihood of successful infiltration. Furthermore, the use of the highly effective zero-day exploits to bypass security measures made containment incredibly difficult. While the initial infection leveraged USB drives, its subsequent spread heavily relied on network-based transmission, indicating that the attackers understood the inherent vulnerabilities within interconnected systems. The malware's creators understood that a single point of entry is never enough for a truly effective attack—multi-vector infections offer redundancy and increase the chance of a successful compromise. The creators expertly planned for the possibility of detection, and incorporated self-preservation and escape mechanisms into the worm's code.

The impact of Stuxnet extended far beyond its immediate target. It demonstrated that nation-state actors possessed the capability to conduct highly targeted, sophisticated cyberattacks capable of causing significant physical damage. The revelation that a piece of malware could not only steal data but also physically destroy industrial equipment altered the landscape of cybersecurity significantly. The subsequent analysis of Stuxnet provided valuable insights into the techniques and capabilities of advanced persistent threats (APTs), prompting governments and organizations worldwide to re-evaluate their cybersecurity posture and allocate more resources to the defense of critical infrastructure. The sheer precision and complexity of Stuxnet's design spurred substantial advancements in the development of improved security protocols for ICS environments.

Researchers painstakingly reverse-engineered Stuxnet, dissecting its code to understand its functionality and propagation mechanisms. This analysis revealed numerous innovative techniques, including the use of rootkits to maintain persistence, polymorphism to evade detection, and the exploitation of previously unknown vulnerabilities. The detailed study of Stuxnet's architecture, including its various modules and their functionalities, provided crucial insights for developing more effective detection and prevention methods. The knowledge gained from this analysis has been instrumental in enhancing cybersecurity defenses and improving the overall security posture of various industrial control systems.

The long-term consequences of the Stuxnet attack remain significant. It significantly raised the bar for malware sophistication, setting a new benchmark for advanced persistent threats. The implications for critical infrastructure security are profound, underscoring the urgent need for robust security measures to protect essential systems from similar, targeted attacks. The attack highlighted the importance of security awareness within organizational environments, recognizing that human error continues to represent a critical vulnerability point. In addition, the incident underscored the requirement for better cross-sector collaboration and information sharing to mitigate the ever-evolving threat landscape.

The Stuxnet case remains a powerful illustration of the evolving threat landscape. Its sophisticated techniques and devastating impact serve as a stark reminder of the potential consequences of advanced cyberattacks. Understanding Stuxnet's architecture and capabilities is crucial for developing effective countermeasures and enhancing the security of industrial control systems and other critical infrastructures. Its influence extends far beyond the immediate consequences of its specific mission, affecting the very way we conceptualize cybersecurity at the highest levels. The legacy of Stuxnet continues to shape the development of both defensive and offensive cybersecurity practices, creating a new era of cyber warfare. The lessons learned from this watershed event continue to resonate throughout the global cybersecurity community.

6.2 NotPetya's Impact

The NotPetya attack, launched in June 2017, wasn't your typical ransomware operation. Disguised as a tax software update, it rapidly spread globally, crippling businesses and causing billions of dollars in damage. Unlike many ransomware strains focused solely on encryption, NotPetya aimed for complete system destruction. Its creators demonstrated a level of sophistication rarely seen, exploiting a vulnerability in the widely used accounting software, M.E.Doc, to infiltrate networks. This wasn't a simple case of encrypting files and demanding a ransom; NotPetya's primary goal was to inflict maximum disruption, showcasing a malicious intent far exceeding financial gain.

Its propagation mechanism relied on a combination of sophisticated techniques, including self-replication and the exploitation of EternalBlue, an exploit leaked from the National Security Agency (NSA). This exploit allowed NotPetya to spread laterally through networks, infecting machines without requiring user interaction. The speed and efficiency of its spread were staggering, highlighting the vulnerability of interconnected systems. Many organizations, believing they had robust cybersecurity measures in place, found themselves completely paralyzed within hours. This underscored the critical need for a comprehensive and proactive security posture, extending beyond traditional antivirus solutions.

The impact extended far beyond immediate financial losses. Supply chains were severely disrupted, production lines halted, and critical services interrupted. Global corporations like Maersk, FedEx, and Merck experienced significant operational setbacks, incurring massive costs in recovery and remediation. The sheer scale of the disruption demonstrated the far-reaching consequences of a successful cyberattack, affecting not just individual organizations but entire industries. The ripple effect emphasized the interconnectedness of the modern global economy and its vulnerability to sophisticated cyber threats.

What truly distinguished NotPetya was its lack of a typical ransomware component. While some encryption did occur, the primary focus was on data destruction. This eliminated the possibility of a simple recovery through ransom payment, making the attack exceptionally destructive. The perpetrators' objectives remain unclear, with theories ranging from espionage to simple chaos creation. Regardless of the motivations, the attack served as a potent reminder of the devastating potential of advanced malware and the importance of preparedness.

Recovery proved exceptionally challenging. Simply restoring data from backups often proved insufficient, as the malware had overwritten system files and master boot records, rendering backups unusable. Organizations were forced to rebuild systems from scratch, a costly and time-consuming process. This underscored the crucial role of robust, multi-layered backup strategies, including offline backups and air-gapped systems, to protect against sophisticated attacks that bypass conventional security measures. The attack exposed critical gaps in many organizations' security architectures, revealing a widespread vulnerability to advanced persistent threats.

Beyond the immediate damage, NotPetya left a lasting impact on the cybersecurity landscape. It forced organizations to reassess their security practices, prompting increased investment in threat intelligence, endpoint detection and response (EDR) solutions, and security awareness training. The attack highlighted the importance of proactive threat hunting, vulnerability management, and incident response planning. Moreover, it intensified the debate surrounding the responsible disclosure of vulnerabilities and the potential consequences of cybersecurity tools falling into the wrong hands. The long-term effects of NotPetya continue to shape the cybersecurity strategies of organizations worldwide.

In the aftermath, many organizations adopted a more proactive approach to security, focusing on threat intelligence, vulnerability management, and incident response planning. They moved towards more robust backup and recovery strategies, implementing measures like air-gapped backups and enhanced data protection mechanisms. The attack emphasized the critical need to go beyond reactive security measures, adopting a proactive approach to identify and mitigate potential threats before they can inflict damage. The lesson learned wasn't just about technological improvements but also about a fundamental shift in mindset, from reacting to threats to actively preventing them.

The NotPetya attack serves as a cautionary tale. It's a stark reminder that even the most sophisticated organizations can fall victim to advanced malware. It was a wake-up call to the entire global community, demonstrating the destructive power of well-crafted malware and the necessity of robust cybersecurity practices. The long-term consequences of NotPetya continue to influence cybersecurity strategies and highlight the need for ongoing vigilance and adaptation in the face of ever-evolving threats. The lessons learned from this devastating attack remain highly relevant today, emphasizing the need for continuous improvement in cybersecurity practices. Investing in robust security measures, developing comprehensive incident response plans, and fostering a culture of security awareness are essential to mitigating the risk of similar catastrophic events in the future.

6.3 WannaCry's Global Reach

The WannaCry ransomware attack, erupting in May 2017, wasn't just another cybersecurity incident; it was a stark demonstration of the global interconnectedness of digital infrastructure and the devastating potential of a single, cleverly designed piece of malware. Exploiting a vulnerability in older versions of Microsoft's Windows operating system, WannaCry spread with alarming speed, crippling hospitals, businesses, and even transportation networks across the globe. Its rapid propagation highlighted the urgent need for robust patching practices and proactive cybersecurity measures, lessons learned the hard way by numerous organizations. The sheer scale of the outbreak, affecting over 230,000 computers in more than 150 countries, underscored the vulnerability of systems reliant on outdated software and the critical need for timely updates.

Unlike previous ransomware attacks, WannaCry's "worm-like" capabilities allowed it to self-propagate across networks without requiring user interaction, significantly expanding its reach and impact. This self-replicating characteristic amplified the threat exponentially, transforming a localized incident into a global crisis within a matter of hours. The rapid spread shocked many, demonstrating how quickly even seemingly contained vulnerabilities can escalate into worldwide disruptions. The attack's efficiency served as a sobering reminder of the interconnected nature of modern infrastructure. Systems were compromised not just because of outdated software, but also because of networked environments. This network effect meant that a single vulnerable machine could act as a springboard for an attack to spread throughout entire organisations. Many companies found the virus spreading unchecked from branch office to head office in a matter of minutes.

The financial and operational consequences of the WannaCry attack were immense. Hospitals were forced to divert ambulances, delaying critical care for patients. Businesses experienced significant downtime, leading to substantial financial losses and disrupted operations. This disruption wasn't limited to large corporations, small and medium-sized enterprises (SMEs) also suffered significantly; many lacked the resources to recover quickly from the attack, leading to long-term consequences. The global impact, measured in billions of dollars in damages, served as a stark reminder of the economic repercussions of such wide-spread cyberattacks. The attack exposed the considerable vulnerability present in crucial industries across nations.

While the ransom demand itself was relatively small, the real cost was the disruption, data loss, and the immense effort required for recovery. Many victims chose not to pay the ransom, a wise decision given the lack of guarantee of data recovery, and the ethical implications of rewarding malicious actors. The attack revealed a crucial gap in the collective response to cyber threats: a lack of global coordination and information sharing. A more unified approach, fostering enhanced collaboration between governments, businesses, and cybersecurity professionals, could have significantly mitigated the impact. It also highlighted that no single entity, however well-prepared, is immune to such large-scale, network-based attacks.

Investigating the attack's origins and the actors behind it revealed a complex web of geopolitical considerations and criminal motivations. While attributed to the Lazarus Group, a North Korean state-sponsored hacking group, the subsequent spread demonstrated the ease with which malicious code can be replicated and disseminated, even beyond the initial intentions of its creators. WannaCry's unexpected reach underscored the fact that malware's impact is not solely determined by the initial perpetrators but also by the vulnerabilities it exploits and the lack of widespread, robust security measures. The event served as a critical wake-up call; highlighting not just the technological vulnerabilities, but the gaps in collective global cybersecurity preparedness.

6.4 SolarWinds Supply Chain Attack

The SolarWinds attack, unfolding in late 2020, wasn't just another intrusion; it represented a chilling escalation in sophistication. It wasn't a random script kiddie exploit; this was a meticulously planned, state-sponsored operation targeting a critical infrastructure provider, SolarWinds, and its vast customer base. The attackers, believed to be linked to the Russian foreign intelligence service, SVR, didn't target individual machines. Instead, they embedded malicious code directly into the SolarWinds Orion platform update, a piece of software used by thousands of organizations worldwide. This update, seemingly benign, contained a backdoor, cleverly disguised within the legitimate code, allowing attackers persistent and undetected access.

This seemingly simple act of injecting malicious code into a trusted software update had devastating consequences. Imagine the effect: the attackers gained access not through a single vulnerable point but through a trusted vendor. They essentially created a network of compromised systems, a digital Trojan horse army, waiting for their command. The scale was unprecedented, with thousands of victims unknowingly harboring the malicious code, providing a potent platform for espionage and sabotage. The compromise went unnoticed for months, highlighting the critical need for robust security monitoring and detection capabilities. The attackers' patience and meticulous planning showcased the evolution of state-sponsored attacks, moving beyond simple exploits to large-scale, long-term infiltration strategies.

The impact reverberated throughout numerous sectors. Government agencies, technology companies, and critical infrastructure providers were all affected. The sheer breadth of the compromise highlighted the vulnerability of supply chains, demonstrating how a single compromised vendor could cascade damage across an entire ecosystem. The subsequent investigations revealed the attackers' capabilities, including their ability to blend into legitimate network traffic, evade detection, and maintain persistent access for months without detection. They moved laterally across the compromised networks with ease, demonstrating a high level of expertise and mastery of offensive cybersecurity techniques. The attack served as a stark reminder of the fragility of even the most secure networks when faced with a sophisticated and determined adversary.

The SolarWinds attack demonstrated the inadequacy of traditional security measures against advanced persistent threats (APTs). Firewalls and intrusion detection systems, while crucial, proved insufficient against an adversary capable of injecting malicious code directly into trusted software. The attackers' ability to blend into legitimate traffic highlighted the limitations of signature-based detection methods. The attack underscored the urgent need for more robust security measures, including enhanced software supply chain security practices, improved vulnerability management, and the adoption of advanced threat detection capabilities capable of identifying anomalous behavior. This included advanced threat detection technologies and proactive measures that go beyond traditional signature-based approaches.

What can organizations learn from the SolarWinds attack? Firstly, prioritize rigorous software supply chain security. This involves carefully vetting third-party vendors, conducting thorough security assessments of their products and practices, and implementing processes to detect and respond to compromised software updates. Secondly, enhance security monitoring capabilities. Implementing comprehensive security information and event management (SIEM) systems, along with advanced threat detection solutions, is crucial for identifying and responding to sophisticated attacks. These systems should go beyond simple log analysis and focus on detecting unusual patterns and behaviors. Thirdly, invest in security awareness training. Educating employees about the risks associated with phishing and social engineering attacks, and reinforcing good cybersecurity hygiene practices, remains a vital line of defense. The attack highlighted that even the best technical security measures can be circumvented if employees are easily tricked.

The SolarWinds supply chain attack was a watershed moment in cybersecurity, highlighting the growing sophistication and reach of state-sponsored attacks. It demonstrated the vulnerability of supply chains and the limitations of traditional security measures. The lessons learned from this attack are crucial for organizations worldwide. Proactive measures, including enhanced supply chain security practices, improved threat detection capabilities, and comprehensive security awareness training, are no longer optional but essential for protecting against future attacks of this scale and sophistication. The widespread impact serves as a stark reminder of the continuous evolution of cyber threats and the necessity of constantly adapting security strategies. The attack fundamentally changed the way security professionals view supply chain risks, shifting the focus towards proactive, multi-layered defenses rather than relying solely on reactive measures. The future of cybersecurity depends on adapting to these evolving threats and embracing new security paradigms that address the complexities of the modern attack landscape.

6.5 Colonial Pipeline Ransomware

The Colonial Pipeline ransomware attack, unfolding in May 2021, serves as a stark reminder of the devastating consequences of sophisticated cyberattacks targeting critical infrastructure. The attackers, a group known as DarkSide, leveraged a relatively simple but highly effective tactic: exploiting a known vulnerability to gain initial access, then deploying ransomware to encrypt the pipeline's operational systems. This wasn't some complex, multi-stage attack involving zero-day exploits; its success hinged on the pipeline's inadequate security posture. The attackers didn't need to be exceptionally skilled hackers; they merely needed to find an exposed entry point. This highlights a crucial point for organizations of all sizes: fundamental security practices often outweigh complex technological defenses.

The immediate impact was crippling. The pipeline, responsible for supplying nearly half of the East Coast's fuel, was shut down, triggering widespread panic buying, fuel shortages, and economic disruption. The ripple effects extended far beyond the initial disruption, impacting everything from air travel to everyday commutes. This wasn't merely an inconvenience; it was a significant blow to national security and economic stability. The attack's success wasn't a testament to DarkSide's technical prowess alone, but a reflection of the vulnerability of critical infrastructure to even relatively basic attacks.

DarkSide's motives, while financially driven, extended beyond mere monetary gain. The attack demonstrated their capacity to disrupt essential services, highlighting the potential for such attacks to be leveraged for geopolitical purposes. This isn't a purely hypothetical concern; the increasing interconnectedness of critical infrastructure and the sophistication of cyberattackers create a significant risk of future attacks targeting essential services. Understanding these motivations, however, necessitates a move beyond simple financial gain, recognizing the political and strategic dimensions at play.

The response to the attack was multifaceted, highlighting the complexities involved in dealing with such events. The FBI was involved, working to trace the attackers and recover the encrypted data. Colonial Pipeline ultimately paid a $4.4 million ransom, a decision that sparked intense debate regarding the ethics of paying ransoms to cybercriminals. This incident raised critical questions about the effectiveness of existing cybersecurity measures and the need for more robust strategies to protect critical infrastructure. Furthermore, this decision fueled ongoing discussions about the best response to ransomware attacks, emphasizing the delicate balance between mitigating immediate losses and setting dangerous precedents.

What can we learn from the Colonial Pipeline attack? Primarily, it underscores the importance of robust security practices, encompassing everything from basic patching and vulnerability management to comprehensive security awareness training for all personnel. A layered security approach, incorporating multiple defenses and regular security audits, is essential. Investing in advanced threat detection and response systems is crucial, enabling organizations to detect and respond to malicious activity more swiftly and effectively. But it's not solely about technology. Human error remains a significant vulnerability, making continuous security awareness training a non-negotiable aspect of any comprehensive security strategy.

The Colonial Pipeline ransomware attack wasn't about deploying some groundbreaking, never-before-seen malware. It was about exploiting a fundamental weakness, a lapse in basic security protocols. The attackers weren't masters of some esoteric hacking technique. They simply took advantage of an opportunity that should never have existed. This reinforces the need for a holistic security approach. The focus shouldn't solely be on the latest and greatest technologies, but on consistently implementing foundational security principles. This holistic approach, integrating technological safeguards with effective security training and protocols, is crucial for preventing future attacks and mitigating their impact. Regular security assessments, testing, and updates are non-negotiable steps for ensuring resilience against these kinds of threats.

Ultimately, the Colonial Pipeline attack should serve as a wake-up call. It's not just about protecting corporate networks; it's about safeguarding critical infrastructure that underpins our daily lives. The consequences of failure are far-reaching and profoundly impact not only the victims but society as a whole. The lesson is clear: proactive, multi-layered security, coupled with a focus on basic security practices and continuous employee training, is crucial for mitigating the risk of ransomware attacks and preventing another such catastrophic event. The cost of inaction far outweighs the cost of proactive security measures. This is a lesson that must be learned, not just by businesses, but by government agencies and all those responsible for maintaining our critical infrastructure.

7 Future of Malware and Cybersecurity

The relentless arms race between malware creators and cybersecurity professionals continues, driven by technological innovation and the ever-increasing value of digital assets. We've seen malware evolve from simple self-replicating programs to sophisticated, multi-stage attacks capable of evading even the most robust security measures. The future holds both daunting challenges and exciting opportunities for those committed to defending against the next generation of threats.

One of the most significant shifts will be the increased reliance on artificial intelligence and machine learning. While AI is already being used for malware detection, its capabilities will dramatically expand, leading to more proactive and adaptive security systems. Imagine AI systems capable of predicting emerging attack vectors, identifying zero-day exploits before they're widely deployed, and automatically responding to threats in real-time, significantly reducing response times and minimizing damage. However, this also presents a counterpoint; malware authors will inevitably leverage AI to develop more cunning and evasive malware, creating a continuous feedback loop of innovation on both sides.

The rise of quantum computing poses a particularly challenging threat. Quantum computers possess the theoretical power to break widely used encryption algorithms, rendering current security protocols obsolete. This potential breakthrough necessitates a fundamental shift in cryptographic approaches, a move towards quantum-resistant algorithms that can withstand the computational power of these emerging machines. The transition will be complex, requiring significant investment in research, development, and widespread adoption of new standards across all sectors. It's a long-term strategy, demanding proactive collaboration among governments, researchers, and private companies.

Beyond technological advancements, the human element remains critical. Sophisticated social engineering techniques, like highly targeted phishing campaigns and carefully crafted disinformation, continue to be highly effective in bypassing technological security measures. The human factor is the weak link in many security systems. Therefore, robust security awareness training will remain paramount. This means moving beyond simple awareness programs to develop a cybersecurity culture within organizations, fostering critical thinking and promoting secure practices at every level. Regular training, combined with simulated phishing attacks and incident response drills, will be vital in strengthening the human element of cybersecurity.

The Internet of Things (IoT) explosion presents another significant challenge. The sheer volume and diversity of connected devices create an exponentially larger attack surface. Many IoT devices lack robust security features, making them easy targets for malicious actors. Securing the IoT landscape will require collaborative efforts from manufacturers, developers, and policymakers to develop standardized security protocols and encourage the adoption of secure-by-design principles. This means integrating security considerations from the initial stages of device development, rather than treating security as an afterthought.

Furthermore, the increasing reliance on cloud services and remote work presents unique vulnerabilities. Protecting data and systems in distributed environments necessitates sophisticated access control mechanisms, robust data encryption, and continuous monitoring. The adoption of cloud-native security tools and practices is no longer optional; it's a necessity. A multi-layered security approach that includes network security, endpoint protection, and data loss prevention will be crucial in mitigating risks.

In conclusion, the future of malware and cybersecurity is a dynamic landscape shaped by technological progress and human behavior. While the challenges are substantial, the development of advanced AI-driven security systems and the adoption of quantum-resistant cryptography offer promising solutions. Ultimately, success relies on a multifaceted approach that encompasses technological innovation, robust security practices, and a commitment to fostering a strong cybersecurity culture across all sectors. The race to protect against future threats is an ongoing endeavor, demanding continuous adaptation and collaboration to stay ahead of the evolving landscape of cybercrime.

7.1 AI in Malware Detection

The increasing sophistication of malware necessitates a paradigm shift in detection strategies. Traditional signature-based approaches, relying on identifying known malware patterns, are proving increasingly ineffective against polymorphic and metamorphic threats that constantly alter their code. This is where artificial intelligence (AI), specifically machine learning (ML), emerges as a powerful countermeasure, offering the potential to identify malicious code even in its novel forms.

ML algorithms, unlike their signature-based counterparts, learn from vast datasets of known malware samples, identifying subtle patterns and characteristics that indicate malicious behavior. This ability to learn from data allows AI systems to adapt to the ever-evolving tactics of malware authors, recognizing even zero-day exploits – malware never encountered before. Instead of relying on predefined signatures, ML models analyze features like code structure, system calls, network traffic, and registry modifications to build a probabilistic model of malicious activity. This approach allows for the detection of previously unknown malware variants, significantly improving security postures.

One effective application of AI in malware detection is through anomaly detection. By training an ML model on the "normal" behavior of a system, deviations from this baseline can be flagged as potential threats. This proves particularly useful in identifying stealthy malware that seeks to blend into the background noise of system activities. Imagine, for example, a network monitoring system employing an AI-driven anomaly detection algorithm. It analyzes network traffic, identifying unusual patterns that suggest unauthorized data exfiltration or malicious communication, potentially thwarting a data breach before significant damage occurs. This capability extends beyond network analysis; it can be implemented within endpoint security software to detect malicious processes and suspicious file activities on individual devices.

Another crucial application of AI lies in the automation of malware analysis. Manually analyzing large volumes of potentially malicious files is both time-consuming and resource-intensive. AI-powered tools can streamline this process, automatically classifying files as benign or malicious based on their characteristics. This automated classification frees up human analysts to focus on more complex cases, requiring detailed manual investigation and analysis. Furthermore, AI can assist in prioritizing malware samples for analysis, focusing on the most suspicious and potentially damaging ones. This prioritisation ensures that resources are efficiently allocated to address the most critical threats first.

However, the integration of AI into malware detection isn't without its challenges. Adversarial machine learning, a field dedicated to manipulating ML models, poses a significant threat. Malware authors might deliberately design their code to evade AI-based detection systems, employing techniques to obfuscate malicious behavior and manipulate the features used by the ML models. Moreover, the effectiveness of AI models is heavily reliant on the quality and size of the training data. A biased or insufficient dataset can lead to inaccurate classifications and reduced effectiveness. Therefore, constant model retraining and adaptation are crucial to maintaining accuracy in the face of evolving malware tactics and to ensure the robustness and resilience of the AI-driven security systems. The ongoing arms race between AI-powered detection systems and adversarial malware necessitates continuous research and development to stay ahead of the curve and build robust defensive strategies.

Finally, the ethical implications of AI in cybersecurity also deserve attention. The potential for misuse of AI-powered tools, such as for mass surveillance or targeted attacks, necessitates careful consideration. The development and deployment of AI-based security systems should be guided by principles of transparency, accountability, and respect for privacy. A responsible approach, considering both the benefits and risks associated with AI, is critical for ensuring its effective and ethical integration into our security infrastructure. The future of cybersecurity will likely involve a collaboration between humans and AI, leveraging the strengths of both to combat the ever-growing sophistication of malware threats. A well-trained and supported human analyst remains a vital component of a robust cybersecurity system, ensuring that AI systems are utilized effectively and that their limitations are well-understood and mitigated.

7.2 The Threat of Quantum Computing

Quantum computing, still largely theoretical for widespread application, poses a profound threat to the future of cybersecurity, specifically challenging the very foundations upon which much of our current network security rests. The exponential processing power promised by quantum computers could break many of the widely used cryptographic algorithms that currently protect our sensitive data and online transactions. This isn't a distant, hypothetical concern; the potential for disruption is imminent.

Current encryption methods, including RSA and ECC, rely on computationally complex mathematical problems that are practically impossible for even the most powerful classical computers to solve in a reasonable timeframe. These problems, such as factoring large numbers or solving the discrete logarithm problem, form the bedrock of secure communication and data protection. However, quantum algorithms, such as Shor's algorithm, can efficiently solve these problems, rendering these widely used encryption methods obsolete.

Imagine a world where the confidentiality of sensitive data, from financial transactions to government secrets, is completely vulnerable. This is the reality that quantum computing presents, highlighting the urgent need for proactive adaptation and mitigation strategies within the cybersecurity landscape. The threat is not merely theoretical; active research and development in quantum computing are accelerating, bringing the potential for decryption of currently secure data closer than many believe.

The implications for malware are particularly disturbing. Malicious actors, armed with the computational power of quantum computers, could easily decrypt encrypted data, giving them access to a treasure trove of confidential information. This would not only facilitate large-scale data breaches but would also empower more sophisticated attacks, allowing for the manipulation of critical infrastructure and potentially jeopardizing national security. Existing malware could become exponentially more dangerous, and the development of entirely new types of quantum-resistant malware is a serious concern.

The speed at which quantum computers could decipher encrypted communications would render many current security protocols ineffective, leaving networks exposed to unprecedented levels of attack. This is not simply a matter of upgrading software; it demands a fundamental rethinking of security architecture. Current defenses are built upon assumptions that will no longer hold true in a post-quantum computing world.

While the timeline for the widespread availability of fault-tolerant quantum computers remains uncertain, the potential impact is so significant that proactive measures are essential. Research into post-quantum cryptography (PQC) is underway, exploring alternative algorithms that are resistant to quantum attacks. These algorithms rely on different mathematical problems believed to be intractable even for quantum computers. However, transitioning to PQC is a complex undertaking, requiring careful planning, standardization, and significant infrastructure upgrades. It's not a quick fix; it's a long-term commitment.

The challenge extends beyond cryptography. The sheer processing power of quantum computers could also accelerate the development of more sophisticated malware, capable of adapting and evading detection far more effectively than current threats. This necessitates a holistic approach to cybersecurity, encompassing not only cryptographic enhancements but also improvements in detection and prevention methods. Investing in advanced threat intelligence, improved malware analysis techniques, and robust security protocols is crucial.

Therefore, preparing for the quantum computing era requires a multi-faceted strategy. It involves investing in research and development of post-quantum cryptographic algorithms, implementing rigorous security protocols that are resilient to quantum attacks, and strengthening existing cybersecurity infrastructure. Furthermore, it demands collaboration among researchers, cybersecurity professionals, and government agencies to establish standardized procedures and to ensure the widespread adoption of quantum-resistant technologies. The threat is real, and the implications are far-reaching; proactive planning is no longer a luxury—it's a necessity. The future of network security hinges upon our ability to adapt and evolve alongside this technological revolution. Failure to do so would leave us vulnerable to an unprecedented level of cyberattacks and data breaches, with potentially catastrophic consequences. The sooner we address this threat, the better prepared we will be to navigate the complexities of a post-quantum computing world.

www.ingramcontent.com/pod-product-compliance
Lightning Source LLC
LaVergne TN
LVHW022025200926
848481LV00007B/2543
9798349282911